SAE EDGE™
RESEARCH REPORT

Unsettled Issues in Determining Appropriate Modeling Fidelity for Automated Driving Systems Simulation

Sven Beiker, PhD
Silicon Valley Mobility

EDGE DEVELOPMENT TEAM

Atul Acharya, *A3Ventures*
Michał Antkiewicz, *University of Waterloo*
David Barry, *Multek Technologies*
Thomas Bock, *Samsung*
Krzysztof Czarnecki, *University of Waterloo*
Eric Paul Dennis, PE, *Center for Automotive Research*
Cameron Gieda, *AutonomouStuff*
Peter-Nicholas Gronerth, *fka Silicon Valley*
Qiang Hong, PhD, *Center for Automotive Research*
Stefan Merkl, *TÜV SÜD America*
Chad Partridge, *Metamoto*
Robert Seidl, *Motus Ventures*
John Suh, *Hyundai CRADLE*
John Tintinalli, *SAE International*
Paul Wells, *A3Ventures*

SAE INTERNATIONAL®

Warrendale, Pennsylvania, USA

About the Publisher

SAE International® is a global association of more than 128,000 engineers and related technical experts in the aerospace, automotive, and commercial-vehicle industries. Our core competencies are lifelong learning and voluntary consensus standards development. Visit sae.org

SAE EDGE™ Research Report Disclaimer

SAE EDGE™ Research Reports focus on topics that are dynamic, in which knowledge is incomplete, and which have yet to be standardized. They represent the collective wisdom of a group of experts and serve as a practical guide to the reader in understanding unsettled subject matter. They are not meant to provide a recommended practice or protocol. The experts engaged have contributed their own thoughts and points of view and these are not the positions of the institutions or businesses with which they are affiliated. A professional writer has collectivized their input; there is no one contributor's perspective being advanced but rather that of a community of practitioners. SAE EDGE™ Research Reports are the property of SAE International and SAE alone is responsible for their content.

About This Publication

SAE EDGE™ Research Reports provide state-of-the-art and state-of-the-industry examinations of the most significant topics in mobility engineering. SAE EDGE™ contributors are experts from research, academia and industry who have come together to explore and define the most critical advancements, challenges, and future direction in areas such as vehicle automation, unmanned aircraft, IoT and connectivity, cybersecurity, advanced propulsion, and advanced manufacturing.

Related Resources

SAE MOBILUS® Automated & Connected Knowledge Hub
http://sae.org/Mobilus/Automated

SAE Team

Frank Menchaca, Chief Product Officer
Michael Thompson, Director, Standards, Information and Research Publications
Monica Nogueira, Acquisitions Director
Jill Leonard, Product Manager
William Kucinski, Managing Technical Editor

Printed in USA

ISSN 2640-3536
e-ISSN 2640-3544
ISBN 978-1-4686-0117-6

To purchase bulk quantities, please contact: SAE Customer Service

E-mail: CustomerService@sae.org
Phone: 877-606-7323 (*inside USA and Canada*)
724-776-4970 (*outside USA*)
Fax: 724-776-0790

Visit the SAE International Bookstore at books.sae.org

About the Editor

Sven Beiker, PhD, is the Founder and Managing Director of Silicon Valley Mobility, a mobility consulting and advisory firm that specializes in technical diligence, product roadmaps, and business models for mobility topics. The engagements span startups, investors, and corporations in the mobility and adjacent industries. In addition, Dr. Beiker is a lecturer at the Graduate School of Business at Stanford University where he instructs students on strategies for startups and corporations in the field of automated, connected, electrified, and shared mobility.

With his well over 20 years of experience gained during his tenure at McKinsey & Company, Stanford University, and the BMW Group, Dr. Beiker is dedicated to the future of the automobile and personal mobility. His mission is to improve sustainability, safety, efficiency, and convenience in how consumers use automobiles. He combines perspectives from technology, business, policy, and human factors.

Dr. Beiker also serves in advisory boards of several startups in the mobility space, as an advisor to the German American Chamber of Commerce in San Francisco, and as an advisor/co-editor to the Lecture Notes in Mobility of Springer Science+Business Media. He is a continuing contributor to the SAE EDGE™ Research Report series.

contents

Unsettled Issues in Determining Appropriate Modeling Fidelity for Automated Driving Systems Simulation

Abstract

This SAE EDGE™ Research Report identifies key unsettled issues of interest to the automotive industry regarding the challenges of achieving optimal model fidelity for developing, validating, and verifying vehicles capable of automated driving. Three main issues are outlined that merit immediate interest:

First, assuring that simulation models represent their real-world counterparts, how to quantify simulation model fidelity, and how to assess system risk.

Second, developing a universal simulation model interface and language for verifying, simulating, and calibrating automated driving sensors.

Third, characterizing and determining the different requirements for sensor, vehicle, environment, and human driver models.

SAE EDGE™ Research Reports are preliminary investigations of new technologies. The three technical issues identified in this report need to be discussed in greater depth with the aims of, first, clarifying the scope of the industry-wide alignment needed; second, prioritizing the issues requiring resolution; and, third, creating a plan to generate the necessary frameworks, practices, and protocols.

NOTE: SAE EDGE™ Research Reports are intended to identify and illuminate key issues in emerging, but still unsettled, technologies of interest to the mobility industry. The goal of SAE EDGE™ Research Reports is to stimulate discussion and work in the hope of promoting and speeding resolution of identified issues. SAE EDGE™ Research Reports are not intended to resolve the issues they identify or close any topic to further scrutiny.

SVEN BEIKER, PhD
Silicon Valley Mobility

Edge Development Team

Atul Acharya, *A3Ventures*
Michał Antkiewicz, *University of Waterloo*
David Barry, *Multek Technologies*
Thomas Bock, *Samsung*
Krzysztof Czarnecki, *University of Waterloo*
Eric Paul Dennis, PE, *Center for Automotive Research*
Cameron Gieda, *AutonomouStuff*
Peter-Nicholas Gronerth, *fka Silicon Valley*
Qiang Hong, PhD, *Center for Automotive Research*
Stefan Merkl, *TÜV SÜD America*
Chad Partridge, *Metamoto*
Robert Seidl, *Motus Ventures*
John Suh, *Hyundai CRADLE*
John Tintinalli, *SAE International*
Paul Wells, *A3Ventures*

ISSN 2640-3536

Introduction

State of the Industry

As set forth in the initial SAE EDGE™ Research Report: "Unsettled Topics Concerning Sensors for Automated Road Vehicles," the potential for automated driving systems (ADSs) to reduce traffic congestion, accidents, transit times, and fuel consumption - while enabling personal mobility and new business opportunities - makes automation arguably the most important topic for the future of ground vehicles, if not mobility as a whole [1].

However, the development of these complex systems of systems is still wrought with challenges. The sensing, perception, and decision-making algorithms that enable ADSs to process raw sensor data [2] and operate safely must be tested thoroughly before being deployed. ADS developers use simulated environments to expedite this process, running multiple vehicle tests in virtual environments - a continued shift away from traditional automotive development methods toward that style of development that more closely follows the software development life cycle.

According to Germany's Pegasus Research Project - which aims to establish criteria, tools, methods, and test scenarios for ADS - simulation is foundational for establishing safety; however, "simulated testing situations should be verified and, thus, the simulation method be validated." [3]

With simulation models, it has famously been said that all models are *wrong*, but some are *useful* [4]. A simulation involves a model of the system, a model of the environment, and a model of system usage [5]. Thus, no simulation is entirely *perfect*. However, such a simulation could still be *useful* for its defined purposes. The key to simulation is an *optimal* level of fidelity appropriate for developing, testing, and validating a system based on its requirements.

This requires that the digital models in simulations (i.e., sensor, vehicles, environments, and human drivers) have appropriate levels of detail, or fidelity for systems under consideration. For instance, to simulate the sensors of an ADS inside of a digital test environment, it may be important to consider the individual beams from that sensor's laser scanner. However, that level of detail may be not necessary to simulate an entire traffic scene. To do so would also require significant time and resources, potentially affecting the development of the ADS. As noted in Koopman and Wagner, simulation realism for its own sake is not an efficient use of resources [5]. Determining the appropriate level of fidelity for those models and simulations remains a challenge for ADS developers and is the impetus for this SAE EDGE™ Research Report.

Unsettled Issues in Determining Appropriate Modeling Fidelity for Automated Driving Systems Simulation

One of the biggest challenges that engineers of ADSs encounter is achieving accurate and reliable models (e.g., sensors, vehicles, environments, humans) for building a simulation for product development and real-world verification and validation (V&V). Balance needs to be found between accurately representing the real world an ADS will operate in and the simplifications needed for a model to expedite development, reduce maintenance, enhance reliability, facilitate operation, and shorten run times. Fidelity, or the measure of a model or simulation's realism, refers to the accuracy of representing both the state and behavior of a real-world object, feature, or condition [6].

Establishing high-fidelity models can contribute to simulations that closely represent the live driving conditions of a test track or open road. However, there are some unsettled issues regarding how to achieve optimal model fidelity and how to determine the appropriate fidelity for models to keep development times and costs manageable. The three most critical questions as identified by the experts contributing to this report include

- How do we make sure simulation models (i.e., sensor, vehicle, environment, human models) represent real-world systems/agents and their behaviors?
- How do we get to a universal simulation model interface and language (esp. on a subsystem level such as sensors)?
- How do characteristics and requirements apply differently to models for sensors, vehicles, environments, and human drivers?

Ensuring that Simulation Models Represent Real-World Systems

Quantifying Sensor Models

While it is important that all models (including sensor, vehicle, environment, and human driver models) are modeled with sufficiently high fidelity, sensor models are an important example to discuss. As the enabling mechanism for an ADS, sensor and sensor models are often the central focus of ADS simulations and should be thoroughly measured and understood.

The quantification of sensor models and other component models should cover all relevant parameters simulated by a given model: signal strength level, signal-to-noise ratio, frequency range, etc. But without a universal baseline or standard for developing sensor simulations, quantifying sensor fidelity becomes extremely challenging and self-referential.

An example of a self-referential method of sensor quantification would be an engineer that models an environment including stationary objects (e.g., buildings, trees, etc.), temporarily stationary objects (e.g., parked cars, trash cans, etc.), and moving objects (e.g., vehicles, pedestrians, etc.) and then has the simulated ADS navigate using LiDAR and visual odometry (i.e., camera input).

Alternatively, a high-definition map could be used as the ground truth for what stationary objects ADS sensors should perceive in a simulation model. While the high-definition map serves as a reference point for stationary objects, those objects also exist as high-fidelity models. As the ADS maneuvers throughout the simulation and senses stationary environmental models, it compares them to data provided by the high-definition map. Simply put, sensor perception should be compared to the intended model.

In a highly sophisticated simulation, an ADS equipped with LiDAR, camera, and radar may be tasked with looking at an environment where a human is standing in front of a large tree trunk. While the ADS may know that there should be a tree there because it's established on a high-definition map, if the environment is accurately modeled, the sensor model will be able to determine that another object is in front of the tree and that object may be a temporarily stationary or moving object - a very important role for the ADS.

Ultimately, the pursuit of sensor model fidelity and quantification draws upon holistic simulation fidelity (including vehicle, environment, and human driver models) and the automation system. An example of this is the need for stationary objects to be modeled with a high-enough level of fidelity in order for a sensor model to appropriately separate temporarily stationary objects from moving objects and other traffic participants (pedestrians, bike, motorbikes, animals, etc.).

Sensor Timing

Additionally, it is vital to account for sensor synchronization (in time and space) in real-world sensor systems within simulations. Beyond variance in timing discrimination for sensors, sensors are not often delivering data to communication buses at the same rates of speed. Sensor simulations should account for the arrival of data to the buses, by representing the differing sensor data speeds of real-world sensors. The accuracy of this "temporal fidelity" also needs to be assessed, such as determining whether to measure in milliseconds versus microseconds, including the timing of bus contention and latency.

Incorporating appropriate temporal fidelity into a simulation will help ensure that the vehicle responses developed through simulations will transfer to a real-world ADS.

Accurately Modeling Radar

A current belief is that there are no radar sensor models with fidelity that experts deem necessary in simulation. Radar wave propagation is extremely difficult to model; unlike LiDAR waves, radar waves are much longer, and they refract on everything. Some organizations like ANSYS perform automotive radar simulations using antenna simulators with radar cross-section (RCS) targets as a workaround, versus simulating a truly arbitrary scene with millions of polygons [7]. This creates a potential barrier to modeling a high-fidelity sensor system.

Additional challenges in radar sensor modeling include the fact that most radar sensor models are not developed to the same requirements. This means that eligibility for application in test architectures (such as hardware-in-the-loop (HIL), model-in-the-loop (MIL), software-in-the-loop (SIL), etc.) remains unclear. Furthermore, the lack of test methods and benchmarks hinder an unbiased comparison of models, such that no statement of reliability of sensor simulation may be concluded. For a more detailed discussion, see "Measurements Revealing Challenges in Radar Sensor Modeling for Virtual Validation of Autonomous Driving" [8].

Assessing Uncertainty and Compounding Effects

Engineers must be cognizant of measuring risk and ask "Is system risk decreasing with various sensor models? Are the layered sensor systems providing layered redundancy?" In effect, there is a residual risk with every subsequent level of simulation (i.e., components to larger modules, systems, subsystems) [5].

The aim of sensor fusion is to merge multiple streams of sensor data to reduce the risk of a system (e.g., an ADS) from acting on limited, faulty, or noisy data. Typically, utilizing multiple perception technologies provides robustness to specific failure modes of a single sensor. However, this redundancy needs to be tested in simulation as much as it does in the real world, including how different, complementary sensor models overlap (i.e., Do they actually provide system redundancy and increase safety once layered in the system?) - which in turn relies on sensor fidelity.

For ADS perception, if there is only a radar and a camera, there's a certain level of uncertainty with only two components within the perception system. There is a chance that the radar and the camera may both miss an obstacle or an object. Typically, the addition of another redundant system subsequently reduces the system uncertainty. This is a very important situation to simulate with component models, as it ensures that redundancy has matured in parallel with sensing capabilities.

For example, in a simulation an ADS' camera, LiDAR, and radar sensors will perceive a white car in front of it on a normal day and provide input for the automated system's algorithms, telling the ADS to decelerate or stop. Simulations can be run for each separate sensor system and combinations of layered sensor systems. But does the simulation have enough fidelity to model the same scenario in a foggy environment when camera sensor models may not be able to discern the white vehicle from the fog?

Recommendations

While literature exists for risk modeling and failure analysis of the components, it should also be applied to measure the decreased risk of combining sensor models and to examine the potential of compounding effects on system risk.

One of the most interesting and powerful models currently used in machine learning is the Generative Adversarial Network (GAN). Instead of a setup with one object recognition model either recognizing or failing to recognize an object, a GAN model can run parallel, recognizing objects within the training set and then generating its own objects for system recognition. After repetitive generation/discrimination cycles, the generator will present objects to the discriminator that are so close to the input from engineers that the discriminator is required to determine whether the input is artificial or real.

GANs present an interesting opportunity for ADS sensor-modeling applications. The artificial intelligence (AI) system in an ADS could potentially be trained on a real-world environment, which is then modeled inside a simulation. Inside the simulation, the same AI system would begin to be able to determine whether or not modeled objects and environments were accurate, potentially replacing it with its own models. This method of using training data is currently being investigated by engineers at Waymo [9].

Achieving a Universal Model Interface and Language

It would be valuable for simulation environments to have an interface both inside and outside of a sensor model. When simulation modelers design an environment model, they need to design and account for the geometries and textures of a multitude of stationary, temporarily stationary, and moving objects. While plastic bags that float across the road and dispersive media like snow and dust may seem trivial, their impact on an ADS is of great importance. The method for developing these objects is a challenge, and often designers are unsure of whether those items are part of a density model, statistical, or macro-objects themselves.

Additionally, endless variability in modeling environments between original equipment manufacturers (OEMs) means that no simulations share a baseline. A catalog or material library of those objects and their required modeling precision outlined in a common design language would be valuable for simulation developers. Provided that these standardized assets were optimally developed, this could lead to major reductions in simulation cost and development time. Furthermore, precision modeling guidelines agreed upon by multiple industry entities could ensure thorough and precise modeling, higher fidelity simulations, and the deployment of safer ADSs.

Compounding this is the huge diversity in the fidelity required for a multitude of models.

Simulation Model Interfaces

Some experts who contributed to this report raised concerns over a lack of a universal simulation model interface and language for describing models. For some model developers, the interface is a "polynomial in a black box," for others it's a "description of the physical attributes of the real world." This results in uncertain descriptions of the simulation model for product teams inside of the OEM, the ADS companies building the software center, and any vendor selling the system.

The question of vendor trust also comes into play. Component vendors have the most knowledge regarding what an actual system will do in a simulation and what it will do in the real world. The vendor may also supply the simulation model; however, that simulation may be a standardized model as a sensor vendor is unlikely to write 20 different versions of a model and maintain them for different simulation environments.

To facilitate both the deployment of models and the exchange of models between different users (e.g., sensor suppliers and OEMs), a model's fidelity should be clearly defined and objectively quantified. Currently, for instance, no developer is building physical test apparatuses for sensors to enable physical HIL testing at scale. Developers will need to shift from linear sensor simulations to advanced, three-dimensional operational testing.

Compounding that issue, developers and suppliers are not required to provide standardized information regarding the component (e.g., sensor) they develop. While some developers provide white papers for their components, other suppliers reverse engineer their devices from competing designs. Some suppliers require engineers, for example, to run the collected sensor data through the sensor provider's proprietary algorithms. Additionally, the creating of an analog simulation without access to information regarding a sensor developer's digital signal processor (DSP) requires manually developing analog interfaces for testing them.

Materials, Geometries, Reflectivity

The differences in models and the real-world environments can be monumental, with environmental hazards impacting ADS functions (e.g., traversing over speed bumps and the effect on object tracking, or the effect of weather on a sensor). Field testing often reveals whether objects were modeled with a high-enough fidelity.

Incorrectly modeling fog, for example, may lead the vehicle to read actual fog in the field as something else entirely. Similarly, highly reflective, or retroreflective signage in the real world has been known to confuse autonomous vehicle (AV) sensors, leading them to misinterpret highly reflective objects or malfunction. While open-source feature libraries exist, developing a standard model catalog for these types of objects would help simulation developers ensure a baseline level of model fidelity for ADSs.

Connected to this is the fact that, over time, sensor degradation due to environmental factors may occur, and there is no guarantee that sensors retain effective calibration over time [1]. It is also important to model issues concerning degraded or malfunctioning component models or for cases where an ADS acts erratically or off nominally when it records an object incorrectly. However, this could be complicated as simulations or environments become more and more complex with more actors. This level of model fidelity could lead to improved self-monitoring or system-health sensing - an important aspect missing in vehicle actuation.

Recommendations

The component and component model should come from one source, if a vendor provides a real object, they should also provide a virtual counterpart. To avoid multiple component model iterations between the component (e.g., sensor) vendor and integrator, it is necessary for the providers themselves to develop models that match the characteristics of their own real-world products. This requires that they test component model attributes against real-world attributes. Additionally, standard component models would streamline the integration of models into simulations [10]. This might eventually lead to the creation of Standards Developing Organizations (SDO)-approved standardized models.

Beyond sensor models, a master list or catalog of environmental objects - including their textures, geometries, and reflectivity - may expedite simulation development and reduce unforeseen modeling complexity of high-fidelity, high-importance models (this would include objects like fog, rain, pedestrians, etc.). It is important to note that some of the experts participating in the development of this report suggested that a standardized model catalog may not streamline ADS development and time to market, and instead may be beneficial to certain organizations that focus on providing simulation tools and services.

Characteristics and Requirements for Various Sensor, Vehicles, Environment, and Human Drivers Models

For simulating sensors, specifically a LiDAR sensor, the individual beam from a scanner must be considered when determining how it will operate in rain or in highly reflective environments. However, that level of sensor fidelity isn't required when simulating an entire traffic scene to establish how the automation stack behaves in stop-and-go driving, heavy traffic, or unprotected left-hand turn situations. When modeling an environment, every millimeter of an object may be critical depending on the objective of the simulation.

The fidelity for each model may be further defined by what type of simulation test is being performed. For testing LiDAR, it depends on whether a sensor-level simulation is being performed (which would just test the sensor stack and would require high-fidelity sensor models) or testing in a closed-loop simulation is being performed (where the entire stack is being tested). For localization the sensor may only need to put out enough photons to see five to six key objects to determine where the vehicle is in the environment. For object perception on a highway, the model must be run at a higher spin rate with higher fidelity to see fast-approaching objects.

Alternatively, a simulation for pedestrian avoidance maneuvers might require the ADS to understand not only the pedestrian's position but the direction that the pedestrian is facing (i.e., their face and body orientation), as well as limb movement (i.e., elbow and knee joints). This becomes especially critical considering complications like training an AI system to interpret pedestrians wearing loose, baggy clothing or head coverings (e.g., a hooded sweatshirt).

Simulating Driving Rules

To create a closed-loop simulation, truly synthetic simulations need driver models or "agents" to simulate traffic with a high level of fidelity. A traffic environment with high fidelity gives the ADS an opportunity to navigate a shared roadway according to established traffic rules. There may be an opportunity for ADS developers to qualify driving rules, which could be varied or modified based on region. These rules could potentially be checked to determine if the system is behaving appropriately. One such an approach is defined by NuTonomy's "Rulebook" which allows mathematical definition of driving rules [11], including the case of conflicting rules.

This could become as granular as qualifying road behavior in the United States as a whole, and then deriving normal driving behavior in regional settings like California or New Jersey.

A consortium of automotive OEMS (such as Daimler, Volkswagen, FCA, BMW), Tier 1 suppliers (such as Continental, Aptiv, etc.), and technology vendors (such as Intel Corp, Infineon) have recently published an approach called the "*Safety First for Automated Driving*" (or SaFAD) with 12 principles for establishing safe automated driving, but it is unclear whether that can be used to derive regional driving rules. A challenge may be adding those rules to a simulation that can be certified up the tool chain so that the system behaves correctly in various regions [12].

This would also include modeling a system's response to "bullying" such as tailgating or brake checking from another vehicle.

Quantifying and Rating Systems

Ultimately, ADS developers need a way to determine that their platform does perform better than a human. Currently, individual sensors models are not developed or quantified the same way, making the qualification of the entire system a challenge.

Although quantifying performance is an unsettled concern, performance expectations should not be mandated. A tiered structure would be more conducive to free-market competition (e.g., the vehicle passes a minimum or a higher level based on multiple criteria such as capability, path planning, etc.).

While some of the experts who participated in the development of the report suggested that SDOs could participate by rating or certifying that a vehicle indeed fits a certain criterion established for a particular level of automated driving, others pointed to SAE International's definitions for automated driving levels (J3016): "As a practical matter, it is not possible to describe or specify a complete test or set of tests which can be applied to a given ADS feature to conclusively identify or verify its level of driving automation." [13]

Different Requirements for Model Fidelity Based on Different User Needs and Simulation Cases

Fidelity requirements apply differently to sensor, vehicle, environment, and human driver models depending on what mode of simulation is used.

1. Full-sensor simulation: Full-sensor simulation is required when testing perception algorithms or a full closed-loop system. When testing the overall effectiveness of LiDAR of an object (i.e., a full-sensor simulation), a relatively high level of fidelity is required for both the environment model and sensor model. When testing a detailed simulation of a LiDAR sensor, the ideal way to simulate what the sensor would see would be to have high-fidelity models for it to sense (e.g., the sensor shoots photons at a nearby building model comprised of brick, wood, and glass would receive different intensities of return for brick, wood, and glass). Full-sensor simulation tests should help create boundary conditions for integrated system tests, like object-level simulations.
2. Object-level simulation (above the perception stack): Object-level simulations are used for pathfinding, decision-making, maneuver planning, and making sure that system geolocation is correct based on the high-definition map. They do not involve the system's perception stack (i.e., no calculations are being performed to determine if an object is a pedestrian, dog, or parked car). The assumption during object-level simulations is that objects are already known to be correct, thus negating the need for high model perception fidelity (e.g., the perception stack knows that building is located eight degrees to the right and 120 feet away and there is a pedestrian with a dog on a leash crossing the street 20 feet in front of the vehicle and will make path planning and object avoidance maneuvers if necessary.)

A categorization tool for assessing simulation model fidelity is proposed in Table 1.

TABLE 1. Proposed tool for categorizing and assessing simulation model fidelity: test article may include items such as the entire ADS, software, or individual components, and test methods can include full simulations and mixed reality tests.

AV Stack				
		Operating design domain (ODD)	Test article	Test method
Simulation model	Sensor			
	Vehicle			
	Environment			

FIGURE 1. **ADS architecture stack overview.**

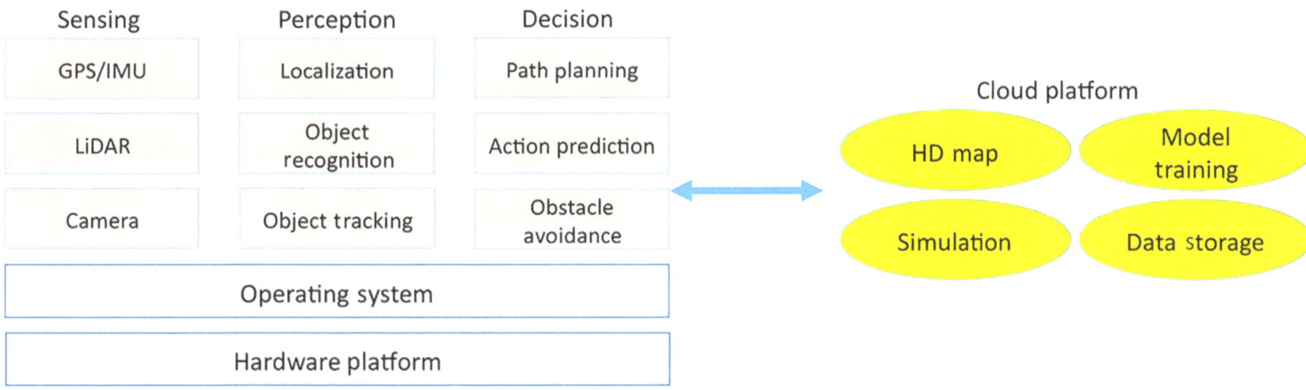

When performing sensor simulations (with the building, pedestrian, and dog), the camera model will take a picture, the LiDAR model will sense, and the data is communicated via the bus architecture of the vehicle. The bus sends that data to the perception stack, objects (such as a parked car) are identified, and then labeled with bounding boxes as needed. The system may then compare this with the high-definition map to confirm that the parked car is a parked car.

Alternatively, an object-level simulation would be a scene of bounding boxes with content labels - the benefit of which would be a huge performance gain for the simulation with far fewer graphics-based calculations. Basically, to simulate something on an object level, the underlying component models need to be established in high fidelity for the same situation. Otherwise, there would be gaps in the simulation framework.

The goal of a model should be carefully stated, including the level and detail of analytical data it is intended to generate. Some models may generate a simple pass/fail checklist, while others may provide detailed performance data for subtle variations in operational conditions. Moreover, various types of simulations require different levels of sensor fidelity. Modeling applications like the effect of weather on sensors needs one level of fidelity, which may involve reflectivity and reduced traction of a road surface - fidelity that may not be necessary in many other applications.

Optimizing Sensor Performance for Human-Machine Interface

There are fundamental differences in developing a system geared for an engineering model and one developed for human perception and real-world efficacy - such as SAE International Level 4 (L4) and Level 5 (L5) ADSs - needs to comply with human expectations of how the AV should operate. Ultimately, the system doesn't need to execute the most mathematically efficient maneuver possible, it needs to maintain passenger trust and comfort. For example, an ADS that swerves violently to avoid an obstacle as soon as possible may cause passenger discomfort (e.g., nausea, panic, etc.). For that situation, it may be more suitable from the user standpoint for the ADS to take less harsh and more gradual avoidance maneuvers.

When a sensor model receives simulated input and interacts with an automation system, which then directs a model of a vehicle to avoid collision with an object, it may not effectively describe what future vehicle occupants might perceive in that situation. Certain types of dynamic vehicle movement and/or repetitive movements could lead to discomfort (e.g., dizziness, nausea, etc.) for vehicle occupants. Furthermore, dynamic and/or repetitive movements could potentially fatigue vehicle components and lead to component failure in extreme scenarios.

But even something as simple as small inconsistent acceleration or slight swerving could lead to uncomfortable occupants who would not purchase the vehicle or utilize it in a ride-sharing service. Sensor and vehicle models should contribute to an understanding of passenger comfort, which in turn affects perceived safety levels and trust.

Different Categories of Simulation that Require Different Levels of Model Fidelity

Managing the various categories of simulation and their various requirements potentially involves breaking the whole simulation down into pieces, from a holistic traffic model to a vehicle model, to the sensor model, etc. Each model will then require engineers to determine the necessary fidelity for each model and layer. All of these combine to additional simulation hours and computing resources. However, uncoupling vehicle dynamics modeling - for instance, from traffic modeling - may have an impact on how sensors are believed to operate in the real world. For example, an automated semitrailer may experience significant vibration while braking, which may render some sensors useless during those maneuvers. Similarly, aggressive control maneuvers in vehicles may potentially cause premature wear on vehicle components or sensors. Even tire dynamics can play into sensor effectiveness, as they are nonlinear. For example, tire slip angles are notoriously difficult to estimate and can add variability in both real-world and simulation maneuvers informed by ADS sensors and simulation models [14].

The question of what models you need in simulation is primarily driven by the operating design domain (ODD) - a specified scenario in which a self-driving vehicle can safely operate on its own (e.g., highway driving). The ODD establishes the requirements and capabilities of the vehicle and can potentially be considered the entire reality of an ADS simulation.

Another consideration is the object under test itself (i.e., what is being tested?). For example, this could be the physical system, software, and individual system capabilities (e.g., maneuvers, path planning, etc.). In normal testing and validation, this manifests as MIL, SIL, HIL, and VIL (vehicle-in-the-loop) testing. Selecting individual system capabilities narrows the scope of testing (into unit testing) and helps determine the level of fidelity needed in a simulation. Both system and unit tests can be accomplished in simulation environments and potentially mixed reality tests, where some aspects are modeled and some components are physical.

Understanding what phenomenon must be modeled and establishing these three factors (the ODD, the test article, and the test method) helps inform the level of fidelity needed for various models in a simulation.

For example, a given ODD establishes an urban driving scenario set in both summer and winter conditions. This automatically informs a simulation developer what kind of general road structure geometry exists, surface textures (e.g., roughness, reflectivity, friction), and roadside infrastructure. For urban driving, horizontal and vertical geometries (e.g., maximum curvature or minimum radius of curves and what types of lane structures are present) and traffic signals are important features that are important to vehicle sensing and operations. Traffic will also be a consideration and could consist of vehicle agents and pedestrian, which require high-fidelity driver behavior models. Finally, the environment will need high fidelity and can be broken out into further categories such as atmospheric environmental objects pertaining to visibility, wind, and lighting and road environmental conditions pertaining to road surface coverage (e.g., ice, snow, rain, mud, etc.) and reflectivity. All of these considerations are derived from the ODD.

Within this example, the vehicle must also be considered (e.g., sedan, minivan, shuttle, etc.) because of different dynamics and performance characteristics and level of autonomy. After establishing these factors, a simulation developer will then need to decide whether to test the entire system of its individual components. If these are being tested in simulation, the entire scope of the simulation will need to be modeled (sensors, vehicle, environment, human drivers) and be verified and validated against the ODD.

There are also certain phenomena that have a big impact over a certain factor. For example, aerodynamics affects driving, but it is relative to speed. Another example may be tire models: at low speeds relatively low-fidelity tire models may be acceptable, but for operation at high speeds with emergency maneuvers, tire models will need to be developed in high fidelity.

Recommendations

Decoupling the stack allows simulations to operate with less modeling fidelity in some circumstances. Traffic simulation can be performed on the object level by giving the middle-of-the-stack objects, without having to do all the low-level processing. This helps reduce cost and development time, but it is only possible if the low-level procession for raw perception and sensor fusion are completed by the time the stack is coupled back together - this includes accounting for sensor-level inputs including things like weather conditions, sensor crosstalk, and decalibration between sensors.

One way to solve the issue involves developing decoupled portions of the stack and modeling applications. Raw perception and sensor fusion need raw data such as sensor level inputs including inputs like weather conditions, sensor crosstalk, and de-calibration between sensors. Traffic simulation can be performed on the object-level by giving the middle-of-the-stack objects, without having to do all the low-level processing. However, this requires that low-level processing will be completed by the time all levels of the stack are coupled back together.

Standardization

Experts contributing to the report identified the development of flight training devices for the aerospace industry as an exemplar for simulation standardization and qualification. International standards are established by organizations like International Civil Aviation Organization and the United States Federal Aviation Administration for how to conduct performance-based qualification of flight simulators [15]. For example, these rules establish quantitative intervals for things like force feedback on a yoke during flight for specific flight conditions or dynamic handling.

Summary/Conclusions

Modeling and simulation address many challenges associated with ADS testing. The number of scenarios and variability in interactions with the environment and objects - and the speed at which these iterative tests can be performed - is integral to testing ADS sensors and automation stacks and assessing system risk. Simulation can also test failure modes and the ADS's responses to those failures [16].

However, creating various simulation models in high fidelity is a significant undertaking - as is determining the priority of which models merit near full fidelity and which can be modeled in lower fidelity. This choice is based on modes, methods, and goals of each simulation. When going to the simulated world - which reflects the ODD - the ADS developer needs to decide which factors can be tested in simulation, and which can be tested in the real world. With unlimited resources, it would be possible to test everything in simulation, but it's not clear that this is cost effective and feasible.

If an ADS developer wants to claim that they have tested maneuvers on slippery roads (for example), their simulation needs to be able to simulate various degrees of friction on the road surface. Ultimately, this determines which parts of the ODD will be verified and validated, which determines which models, factors, and fidelity will be required.

SAE EDGE™ Research Reports

SAE EDGE™ Research Reports, like the present report on "Unsettled Issues in Determining Appropriate Modeling Fidelity for Automated Driving Systems Simulation," are intended to push further out into still unsettled areas of technology of interest to the mobility industry. SAE launches these reports before attempting to form a joint working group, let alone a cooperative research program or a Standards Committee.

SAE EDGE™ reports are intended to be quick, concise overviews of major unsettled areas where vital new technologies are emerging. By definition, an unsettled area is characterized more by confusion and controversy than established order. Early practitioners must confront an absence of agreement. Their challenge is often not to seize the high ground but to find common ground. These scouting reports from the frontiers of investigation are intended merely to begin the process of sorting through critical issues, contributing to a better understanding of key problems, and providing helpful suggestions about possible next steps and avenues of investigation.

SAE EDGE™ Research Reports, therefore, are fundamentally distinct from the more formal working groups approach and far removed from the more mature cooperative research programs and standards' development process. They begin by convening an informal meeting of subject matter experts working in an unsettled area of technology from industry, government, and the research community. A moderated but collegial discussion ensues regarding their shared interest. This conversation attempts to identify critical issues hindering a clear understanding of important topics, preventing a common agreement on potential solutions, or slowing progress in key areas of research.

An SAE EDGE™ Research Report is then drafted and circulated for comment to meeting attendees and other interested parties. After review, the final result of this process is a report like the current examination of unsettled issues concerning the sensors needed for automated vehicles.

Next Steps for Determining Appropriate Modeling Fidelity for ADSs Simulation

This publication should be considered only as a first step toward clarifying the issues modeling fidelity for ADSs simulation. The intention behind this and other SAE EDGE™ Research Reports is to start a dialogue among interested parties on important industry-wide topics that require further attention. The expectation is that these explorations of unsettled areas of technology will lead to the formation of working groups and, ultimately, committees that can address and resolve the issues they raise, producing a framework for developing a common vocabulary of definitions, best practices, protocols, and standards needed to support the continued progress toward safer and more innovative automotive products.

The meetings and interviews that gave rise to this publication demonstrated a great willingness on the part of the industry - and many of its experts - to define simulation fidelity requirements and eventually the standards needed to enable modeling and simulation technology to move ahead as quickly and efficiently as possible. SAE International has demonstrated its lead in this and closely related areas by the introduction of unsettled sensor topics in the SAE EDGE™ Research Report on "Unsettled Topics Concerning Sensors for Automated Road Vehicles," the basic reference work on active safety system sensors (SAE J3088™), the definitions for automated driving levels (SAE J3016™), the first vehicle cybersecurity framework (SAE J3061®), and numerous standards that establish the building blocks of ADAS, electrification, and intravehicular and extravehicular communications.

This SAE EDGE™ Research Report on "Unsettled Issues in Determining Appropriate Modeling Fidelity for Automated Driving Systems Simulation" identifies the following key topics for further pursuit, both through continued informal discussions among industry practitioners and through more formal working groups:

- Assuring various sensor, vehicles, environment, and human drivers models represent the real-world systems they are modeling with appropriate fidelity
- Developing a universal simulation model interface and language
- Characterizing and developing baseline models for sensors, vehicles, environments, and human drivers

Recommendations

The overall recommendations of this SAE EDGE™ report can be summarized as follows:

1. Different categories of simulation require different levels of model fidelity. For example, on the base sensor-level simulations, high fidelity is required to establish a sensor capability. Raw perception and sensor fusion need raw data such as sensor-level inputs including inputs like weather conditions, sensor crosstalk, and de-calibration between sensors. Traffic simulation can be performed on the object level by giving the middle-of-the-stack objects, without having to do all the low-level processing. However, this requires that low-level processing will be completed by the time all levels of the stack are coupled back together.
2. GANs might be applied in parallel with a simulation to recognize objects within the training set and then generate new own objects for system recognition. After repetitive generation/discrimination cycles, the generator will present objects to the discriminator that are so close to the input from engineers that the discriminator is required to determine whether the input is artificial or real. The AI system in an ADS could potentially be trained on a real-world environment, which is then modeled inside a simulation.
3. To avoid multiple component model iterations between the provider and integrator, providers should develop component models that match the characteristics of their real-world products. This requires that they test component model attributes against real-world attributes. Additionally, standard component models would streamline the integration of models into simulations.
4. Creating a standard catalog of environmental objects - including their textures, geometries, and reflectivity - would expedite simulation development and reduce unforeseen modeling complexity of high-fidelity, high-importance models. This would include objects like fog, rain, pedestrians, etc.
5. Refer to the ODD when determining the priority of model fidelity, establishing the requirements and capabilities of the vehicle as well as which models need to be verified and validated in order to properly simulate a test environment.
6. Leverage performance-based qualification practices for flight simulators used in the aerospace industry to inform which systems require high fidelity in automotive simulations.

Abbreviations/Definitions

ADS - Automated Driving System. Intelligent computer control of vehicles based on sensor input, characteristic of vehicles at L3 and higher *Automated Driving Levels.* See the entry for this term below. Automated Driving Levels (as defined by SAE J3016™)

L0 - No Driving Automation. A human driver performs the entire Dynamic Driving Task (DDT). Other vehicle systems may provide warnings or support, e.g., braking systems with automatic skid control.

L1 - Driver Assistance. The ADS is capable of executing either the longitudinal or the lateral vehicle motion control subtask (not both). It disengages immediately upon driver request. A human driver is present at all times to perform all control tasks not done by the automation system, to determine when it is appropriate to engage the automation system, to supervise all operations of the automation system, and to intervene as necessary to maintain safe vehicle operation. Example: automatic speed control ("cruise control") for highway driving.

L2 - Partial Driving Automation. The ADS is capable of executing both the lateral and the longitudinal vehicle motion control subtasks. It disengages immediately upon driver request. A human driver must be present and attentive at all times to perform control tasks not done by the automation system, to determine when it is appropriate to engage the automation system, to supervise all operations of the automation system, and to intervene as necessary to maintain safe vehicle operation. Total driver disengagement from human control mechanisms (hands off the wheel and feet off the pedals) is possible in narrowly limited driving scenarios, e.g., lane centering.

L3 - Conditional Driving Automation. A human driver must be present and attentive at all times to operate the vehicle outside the ODD of the ADS and to take over control of the vehicle from the ADS within its ODD as necessary. The ADS can only be engaged in its ODD but can perform the entire DDT when engaged. It automatically disengages when it determines either that its ODD limits are about to be exceeded or it is experiencing a performance-relevant system failure, first, by issuing a timely request to intervene to the human driver and, second, by disengaging an appropriate time after issuing the request. It also disengages immediately upon driver request. The human driver is responsible for verifying the operational readiness of the vehicle, handling all DDT outside the ODD of the ADS, determining when the ADS can be engaged, and serving as the fallback-ready driver when the ADS is engaged, responsible for maintaining a minimal risk condition by actively intervening in ADS operation or assuming vehicle control on automatic ADS disengagement. An ADS ODD might include, for example, freeway driving or light traffic on open roads outside urban areas.

L4 - High Driving Automation. A human driver is responsible for verifying the operational readiness of the vehicle and determining whether to engage the ADS. The ADS will permit engagement only within its ODD. When the ADS is engaged, the human driver has no fallback role and may ride as a passenger. Once engaged, the ADS will transition to a minimal-risk condition before it disengages 1) as the result of a performance-relevant system failure, 2) after an automatically issued request to intervene is ignored by a human driver, or 3) in response to a user request that it achieves a minimal-risk condition. Optionally, a human driver may reassume DDT, by requesting the ADS disengage and waiting for it to turn over control of the vehicle once the ADS determines that conditions are appropriate; or, in a fallback role, following an ADS request to intervene; or, after the ADS has automatically disengaged, following achievement of a minimal-risk condition.

L5 - Full Driving Automation. The ADS ODD expands to permit engagement under all driver-manageable on-road conditions. L5 vehicle operation, when the ADS is engaged, is identical to L4 vehicle operation.

DSP - Digital Signal Processor

GAN - Generative Adversarial Network

HIL - Hardware In the Loop

LiDAR - Light Detection And Ranging. Parallel term to radar.

MIL - Model In the Loop

ODD - Operating Design (Driving) Domain, a specified scenario in which a self-driving vehicle can safely operate on its own, e.g., highway driving.

OEM - Original Equipment Manufacturer

Radar - Radio detection and ranging. Parallel term to LiDAR.

RCS - Radar Cross Section

SDO - Standards Development Organization

SIL - Software In the Loop

V&V - Verification and Validation

Acknowledgments

Recognition should go first to all the participants, many of who also provided feedback on the draft version of this publication. Without their input and initiative, this SAE EDGE™ Research Report would not have been possible.

Atul Acharya, A3Ventures
Michał Antkiewicz, University of Waterloo
David Barry, Multek Technologies
Thomas Bock, Samsung
Krzysztof Czarnecki, University of Waterloo
Eric Paul Dennis, PE, Center for Automotive Research
Cameron Gieda, AutonomouStuff
Peter-Nicholas Gronerth, fka Silicon Valley
Qiang Hong, PhD, Center for Automotive Research
Stefan Merkl, TÜV SÜD America
Chad Partridge, Metamoto
Robert Seidl, Motus Ventures
John Suh, Hyundai CRADLE
John Tintinalli, SAE International
Paul Wells, A3Ventures

Additional thanks go to the following individuals who also had the opportunity to provide feedback on the draft version of this publication. Their contributions helped shape this publication.

Gagan Aujla
Antenna Group
Gräfe Gunnar
3D Mapping Solutions GmbH

The authors of this document together with the SAE Team responsible for its creation join in expressing our deepest appreciation to all the individuals mentioned above.

Sven Beiker, PhD
Silicon Valley Mobility
(external consultant to SAE International)

Works Cited

1. Beiker, S. "SAE EDGE™ Research Report: Unsettled Topics Concerning Sensors for Automated Road Vehicles," Warrendale, PA: SAE International, 2018.
2. SAE J3088™ Ground Vehicle Standard, "Active Safety System Sensors," Warrendale, PA: SAE International, 2017.
3. Pegasus Research Project, "Subproject 3: Testing," https://www.pegasusprojekt.de/en/subproject-3, accessed September 21, 2019.
4. Box, G., "Robustness in the Strategy of Scientific Model Building," MRC Technical Summary Report #1954, Madison, WI: University of Wisconsin-Madison, 1979.
5. Koopman, P. and Wagner, M., "Toward a Framework for Highly Automated Vehicle Safety Validation," SAE Technical Paper 2018-01-1071, 2018, doi:10.4271/2018-01-1071.
6. Jacquart, J., Voogd, J., Brade, D., and Yi, C., "WEAG THALES JP11.20 (REVVA) Results and Perspectives," 2005 European Simulation Interoperability Workshop, Paper 021, 2005.
7. Carpenter, S., "Autonomous Vehicle Radar: Improving Radar Performance with Simulation," *ANSYS Advantage* XII, 1, 2018.
8. Holder, M. et al., "Measurements Revealing Challenges in Radar Sensor Modeling for Virtual Validation of Autonomous Driving," Maui, HI: IEEE, 2018, https://doi.org/10.1109/ITSC.2018.8569423.
9. Bansal, M., Krizhevsky, A., and Ogale, A., "ChauffeurNet: Learning to Drive by Imitating the Best and Synthesizing the Worst," 2018, arXiv:1812.03079v1, https://arxiv.org/abs/1812.03079, accessed September 10, 2019.

10. Seidl, R., "Benefits of Simulation in the Automated Vehicle Supply Chain," Medium, Published Feb. 28, 2018, https://medium.com/@metamoto/benefits-of-simulation-in-the-automotive-supplychain-985fa2ffcf64, accessed September 11, 2019.
11. Censi, A. et al., "Liability, Ethics, and Culture-Aware Behavior Specification Using Rulebooks," 2019, arXiv:1902.09355v2, https://arxiv.org/abs/1902.09355, accessed September 11, 2019.
12. Intel, "Safety First for Automated Driving," 2019, https://newsroom.intel.com/wp-content/uploads/sites/11/2019/07/Intel-Safety-First-for-Automated-Driving.pdf, accessed September 17, 2019.
13. SAE J3016™ Ground Vehicle Standard, "Taxonomy and Definitions for Terms Related to On-Road Motor Vehicle Automated Driving Systems," Warrendale, PA: SAE International, 2017.
14. Song, S., "Towards Autonomous Driving at the Limit of Friction," Thesis, University of Waterloo, 2014, https://pdfs.semanticscholar.org/d7a6/753a089f9e8fe35871d86b95294ae134aae5.pdf, accessed September 6, 2019.
15. Federal Aviation Administration Advisory Circular, "AC 120-45A - Airplane Flight Training Device Qualification," 1992, https://www.faa.gov/documentLibrary/media/Advisory_Circular/AC_120-45A.pdf, accessed September 10, 2019.
16. National Highway Traffic Safety Administration, "A Framework for Automated Driving System Testable Cases and Scenarios," 2018, https://www.nhtsa.gov/sites/nhtsa.dot.gov/files/documents/13882automateddrivingsystems_092618_v1a_tag.pdf, accessed September 10, 2019.

Contact Information

EDGEresearch@sae.org.

The authors of this document together with the SAE Team responsible for its creation join in expressing our deepest appreciation to all of the individuals who contributed.

www.ingramcontent.com/pod-product-compliance
Lightning Source LLC
LaVergne TN
LVHW070942210926
848482LV00003B/281
9781468601176